GENETICS

by

IAN GRAHAM

WHAT IS GENETICS?

Genetics is a branch of science that deals with the way features such as the colour of our eyes are passed on to us, or inherited, from our parents. Our appearance, some of our behaviour and even some of the illnesses from which we suffer are all determined by our genes, the chemical blueprint for life. By studying these genes, scientists are learning more and more about how our bodies work, and even how to cure many inherited illnesses. Scientists who study genes are called geneticists.

CHARLES DARWIN

The English scientist Charles Darwin (1809–82) was the first person to realize that animals change over many generations. He studied all sorts of animals and plants during a five-year, round-the-world voyage in the 1830s on board the ship *HMS Beagle*. Darwin noticed how the same creatures had changed in different ways to suit the place in which they lived. In 1859, he wrote a famous book called *Origin of Species*. In it, he came up with a theory called evolution to describe the changes he had seen. Darwin could not, however, explain how these changes happened.

SCIENCE EXPLAINED: STUDYING GENETICS

Scientists can study genetics by using the tiniest traces of plant and animal tissue or by looking at whole populations. The people of Iceland are often studied by geneticists, because all Icelanders are descended from one small group of settlers. This enables scientists to study the way all sorts of characteristics have been passed down through many generations of families.

SELECTIVE BREEDING

Farmers have used genetics for thousands of years to change certain characteristics of their animals. By allowing only the finest animals to breed, they made sure that only the best physical characteristics were passed on to the next generation. This has enabled them to breed dairy cows that produce more milk and cattle with more meat. Controlling the way animals breed, instead of letting nature take its course, is called selective breeding.

TIME CAPSULES

Tens of millions of years ago, flies buzzed around the prehistoric world. Sometimes, a fly got trapped inside gooey sap oozing from a tree trunk. The sap set hard and became amber. If the fly's last meal was blood sucked from a dinosaur, then the blood and the genetic information it contains might still be preserved inside the fly's gut. In the film *Jurassic Park*, scientists brought dinosaurs back to life using the samples found in these amber prisons.

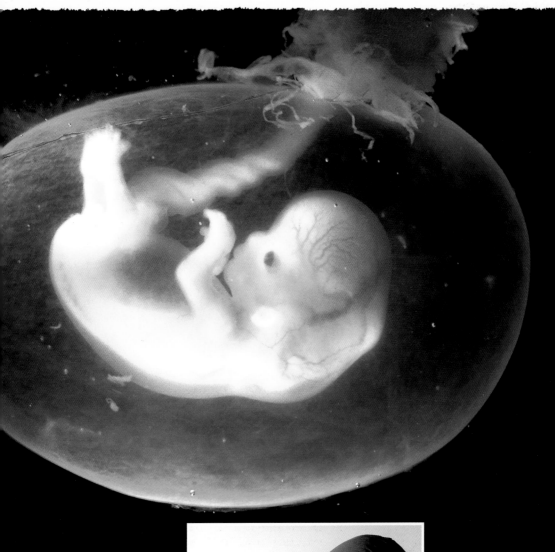

A NEW CRIME FIGHTER

Genetic testing has added a powerful new scientific tool to aid police investigations of crimes. Genetic information can be retrieved from skin, blood and other human tissue found at the scene of a crime. It can be used to eliminate someone from police enquiries or to help secure a conviction. If DNA found at the scene is identical to a sample taken from a suspect, then the chances of a mismatch are literally millions to one.

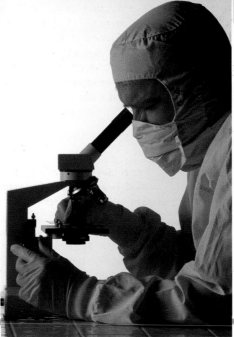

INHERITANCE

Even at an embryonic stage, many of our physical characteristics have already been decided. Information inherited from our parents controls how we develop into a baby, what we will look like at birth and how we will grow into an adult. Even characteristics such as height are inherited and programmed into the developing embryo.

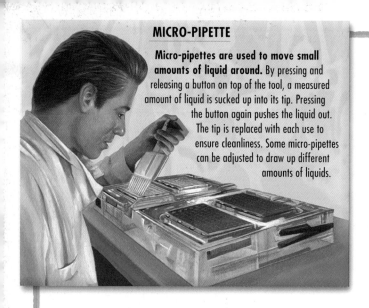

MICRO-PIPETTE

Micro-pipettes are used to move small amounts of liquid around. By pressing and releasing a button on top of the tool, a measured amount of liquid is sucked up into its tip. Pressing the button again pushes the liquid out. The tip is replaced with each use to ensure cleanliness. Some micro-pipettes can be adjusted to draw up different amounts of liquids.

TRICKS & TECHNIQUES

Because geneticists often work with minute pieces of plant and animal material, they need special tools to be able to do their job properly. Optical microscopes are used to look at tiny samples in greater detail, while the development of electron microscopes has enabled geneticists to study the smallest structures inside cells. In addition to visual aids, they use very small tools, or micro-tools, to handle genetic material. Finally, a range of equipment is now available to help geneticists analyse and understand the genetic information their studies uncover.

OPTICAL MICROSCOPE

Optical microscopes (or light microscopes) use lenses to magnify tiny samples up to 1,000 times. They use two or more lenses. One lens, the objective, forms a magnified image of the sample, which is viewed through a second lens, the eyepiece. The magnifying power of the microscope is the magnifying powers of the two lenses multiplied together. So, a microscope with an objective that magnifies 40 times and an eyepiece that magnifies ten times makes samples look 400 times bigger. Microscopes are usually fitted with at least three different objective lenses, so that different magnifying powers can be selected.

ELECTRON MICROSCOPE

An electron microscope can magnify objects up to one million times, showing the structures inside a cell in sharp detail. Instead of light, they use beams of particles called electrons to form an image. The electrons either bounce off the sample or pass through it and then hit a screen, which turns them into light. The image can also be displayed on a computer screen.

DNA SEQUENCER

Scientists who are trying to understand the genetic information in our bodies use a computer-controlled machine called a sequencer. It automatically works out the order, or sequence, of the chemical units that make up our body's genetic code. Knowing this is vital to understanding how our genetic code controls our bodies.

MICRO-TOOLS

Geneticists sometimes want to inject material into a cell, so they have to use tools that are small and gentle enough to hold it without causing any damage.
Extremely fine needles are also needed, with a tip narrow enough to fit inside a single cell. A cell is usually held still by sucking it onto the end of a fine glass tube, like picking up a sweet by sucking it onto the end of a straw.

ELECTROPHORESIS CELL

Scientists use a technique called electrophoresis to compare genetic material from different people.
The job is done by using an electrophoresis cell (*right*). It produces a photograph, called an autoradiograph, that shows a ladder-like pattern of bands for each sample. By looking at how similar or different the patterns are, scientists can tell whether the samples came from people who are related to each other.

SCIENCE EXPLAINED: ELECTROPHORESIS CELLS

Electrophoresis cells are used to compare genetic material. A blob of liquid containing pieces of genetic material is squirted into one end of a sheet of gel. An electric current flowing through the gel from one end to the other makes the genetic material move through the gel. The short pieces travel further than the long pieces because they can move through the gel more easily. A sheet of photographic film is laid on the gel. The genetic material, which has been made radioactive, darkens the film, forming the ladder-like pattern of bands that show how far the various fragments have travelled. The patterns can then be compared with other examples.

WE ARE ALL UNIQUE

There are about six billion people alive in the world today, but it is very rare to see two or more who look the same. So many factors make up our appearance that a tiny change in just one or two of them can distinguish us from each other. Even people who look alike may be very different in ways that we cannot see. They may have different blood groups or be affected by different medical conditions. So, even if two people are similar in appearance, scientists can still tell them apart.

A FACE IN THE CROWD

The thousands of faces in a crowd of people are all roughly the same shape and size. They all have two eyes, two ears, a nose and a mouth, but yet they are all distinctive. The unique characteristics inherited from their parents make them look different from each other. Our brains are very good at spotting these tiny differences.

SCIENCE EXPLAINED: THE GENETIC CODE

We are different from each other because the genetic information, or code, that builds our bodies and tells them how to develop is so incredibly complicated. The chance of two people having exactly the same genetic code (apart from identical twins) is about the same as the chance of two people tossing a coin tens of thousands of times and coming up with exactly the same result.

RETINAL PATTERNS

The retina is the light-sensitive layer at the back of each eye. As an eye develops, blood vessels grow across the retina. The pattern the blood vessels make is unique to each person and can be used for identification purposes. Retinal identification works by firing an invisible infra-red beam into the eye and picking up the reflection of the vessels from the retina. The pattern is then compared to scans already stored in an image bank.

IDENTICAL TWINS

A woman normally produces one egg at a time. If it is fertilized by a man's sperm, it might develop into a baby. However, if the fertilized egg splits into two, each part can develop into a baby, and the babies will be identical to each other. If a woman produces two or more eggs at the same time, then each one has the potential to develop into a baby. But, because they have come from separate eggs, the babies will not be identical to each other.

EYE ON THE FUTURE

Our eyes can be used to identify us because, like fingerprints, they contain unique patterns. The iris, the coloured part of the eye that surrounds the black pupil, is made from a ring of muscle. The patterns and colours in the muscle are unique to each person. Banks are already testing new cash machines that identify people before issuing money to them by scanning their irises and comparing the results to images stored in a computer.

FAMILIES

Look at any family and you'll probably notice that the children often look like at least one of their parents. Sometimes, children look more like their grandparents because some characteristics do not show up in every generation. The tendency to have red hair, for example, might be inherited, but it may be skip a generation before the next person with red hair is born into that family.

FINGERPRINTS

We all have patterns of raised lines on the ends of our fingers. They form before we are born and stay with us throughout our lives. People have marked important documents for thousands of years with their fingerprints. Because everyone's fingerprints are unique, police can use them to identify people.

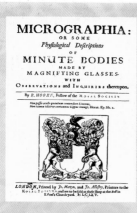

THE CELL PIONEER

The English scientist Robert Hooke (1635-1703) was one of the first people to look at living things through a microscope. He built his own, and in 1665, he published a book, called *Micrographia*, with drawings and descriptions of his findings. Hooke was also the first person to use the word 'cell' to describe the walled bags of material he saw inside living things.

STORY OF A CELL

The genetic material inside the cell's nucleus controls which proteins the cell makes. It makes them from substances that enter the cell through its outer membrane. The proteins are formed in small ball-shaped objects called ribosomes, and stored in folded layers of membranes around the nucleus called the endoplasmic reticulum. Sausage-shaped objects called mitochondria break down molecules to release the energy the cell needs. A folded structure called the Golgi complex stores substances and moves them around to where they are needed.

CELLS & THEIR STRUCTURE

Plants and animals are made from cells. A cell is a microscopic bag of material called protoplasm, which is made from protein. The protoplasm is divided into two parts, called the cytoplasm and the nucleus. It is the nucleus that controls what happens in the rest of the cell. Each cell has a specific purpose. The different parts of plant flowers, roots and leaves, and animal muscle, bone and skin are made from specialized cells. The simplest living things on Earth have just one cell, while the most complex are made from trillions.

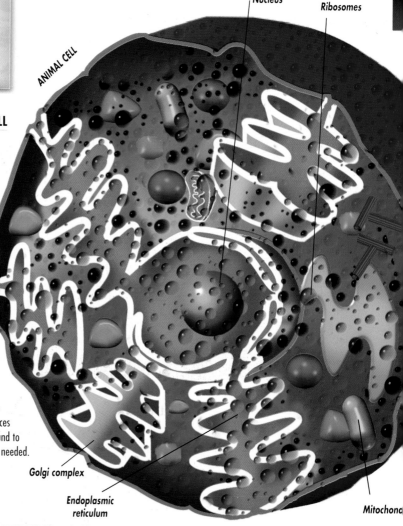

ANIMAL CELL

Nucleus

Ribosomes

Golgi complex

Endoplasmic reticulum

Mitochondia

A SIMPLE LIFE

The simplest organisms, like this *Amoeba proteus*, are made from only one cell, while larger, more complex organisms are made from lots of cells working together. The first cells that appeared on Earth more than three billion years ago had no nucleus. Their DNA was distributed throughout the cytoplasm instead. Cells with a nucleus evolved about a billion years later, and the first organisms with more than one cell appeared about 550 million years ago. Animal cells normally have one nucleus, but red blood cells have no nucleus and this single-celled organism *Amoeba paramecium* has two nuclei.

SCIENCE EXPLAINED: MITOCHONDRIA

Mitochondria are tiny chemical factories inside cells that break down sugars to release energy. They contain genetic material inherited only from the mother. As this was copied down through the generations from mother to daughter, it changed from time to time because of natural mistakes called mutations. By studying these mutations in mitochondria from all sorts of creatures, scientists can tell which creatures evolved from which, in what order and when.

CELL TYPES

Human beings have about 200 different types of cells, including nerve, muscle, bone, skin and blood cells (*right*). They take in food molecules and break them down into smaller, simpler molecules. The cells then use these substances to make proteins that build and repair cells and take part in the many chemical reactions that go on throughout the body.

THE DOUBLE HELIX

DNA is a very long molecule with a shape called a double helix. It is made of
billions of chemical units called nucleotides linked together like beads strung out along a
string. Each nucleotide is made from three parts – a chemical called a phosphate, a sugar
called deoxyribose and a molecule called a base. The phosphates and sugars form the two
long intertwined strands of the double helix and the bases link these two strands together.

MAKING THE BREAKTHROUGH

**The English scientist Francis Crick (1916–)
and the American scientist James Watson
(1928–) discovered the structure of the DNA
molecule in 1953 while they were working
together at Cambridge University in England.**
They won the Nobel Prize in 1962 for their findings.
The discovery of DNA was ground-breaking in enabling
us to understand the make-up of living things.

A CLOSER LOOK AT DNA

Plants and animals grow and function because of the amazingly complicated instructions in the genetic material inside their cells. The genetic material is stored in the form of a substance called DNA (deoxyribonucleic acid). Most DNA is found inside a cell's nucleus. Some DNA and a different sort of nucleic acid called RNA (ribonucleic acid) are found outside the nucleus in the rest of the cell. DNA is a very long molecule that can copy itself so that all the new cells produced by a plant or animal as it grows contain copies of the same set of genetic instructions, or genetic code.

SEX CELLS

There are two types of sex cell, the female ovum (egg) and the male sperm. Both cells contain only half the DNA found in other cells. When an egg is fertilized by a sperm, they join together and their DNA combine. Together, they form a new cell with the correct amount of DNA.

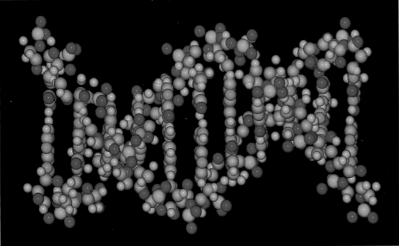

COILS OF COILS

The DNA molecule (*left*) is twisted round itself to form a coil, like a piece of string wrapped around a pencil. That coil is then twisted round itself again to form a coiled coil. If the coiled coil of DNA in just one human cell was unwound and stretched out in a straight line, it would be more than 3 metres (10 ft) long. If all the DNA in your body was unravelled, it could stretch to the sun and back 600 times!

AN EARLY PIONEER

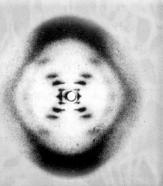

The key breakthrough that enabled Watson and Crick (see far left) to discover the double helix structure of DNA was provided by another scientist, Rosalind Franklin (1919–1958). She had produced patterns on photographic film by firing X-rays at DNA. X-rays normally travel in straight lines, but DNA bends them. The bent X-rays make a pattern of spots on the film (*right*). These patterns gave Watson and Crick vital clues to the structure of DNA.

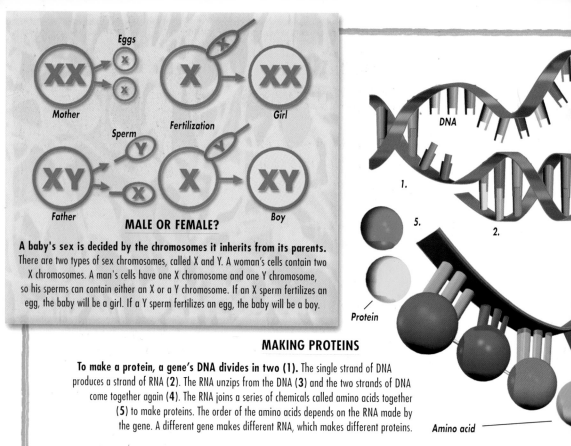

MALE OR FEMALE?

A baby's sex is decided by the chromosomes it inherits from its parents. There are two types of sex chromosomes, called X and Y. A woman's cells contain two X chromosomes. A man's cells have one X chromosome and one Y chromosome, so his sperms can contain either an X or a Y chromosome. If an X sperm fertilizes an egg, the baby will be a girl. If a Y sperm fertilizes an egg, the baby will be a boy.

MAKING PROTEINS

To make a protein, a gene's DNA divides in two (1). The single strand of DNA produces a strand of RNA (**2**). The RNA unzips from the DNA (**3**) and the two strands of DNA come together again (**4**). The RNA joins a series of chemicals called amino acids together (**5**) to make proteins. The order of the amino acids depends on the RNA made by the gene. A different gene makes different RNA, which makes different proteins.

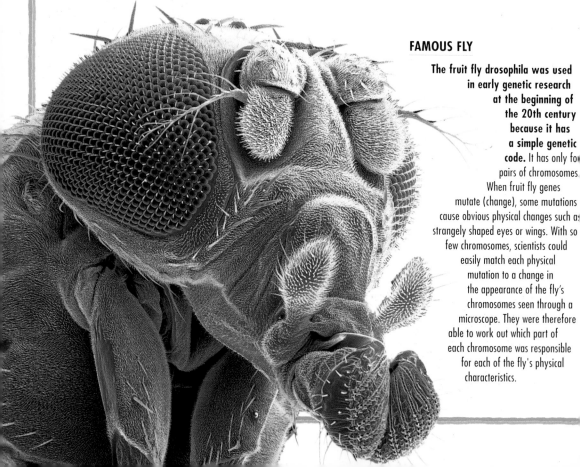

FAMOUS FLY

The fruit fly drosophila was used in early genetic research at the beginning of the 20th century because it has **a simple genetic code.** It has only four pairs of chromosomes. When fruit fly genes mutate (change), some mutations cause obvious physical changes such as strangely shaped eyes or wings. With so few chromosomes, scientists could easily match each physical mutation to a change in the appearance of the fly's chromosomes seen through a microscope. They were therefore able to work out which part of each chromosome was responsible for each of the fly's physical characteristics.

GENES & CHROMOSOMES

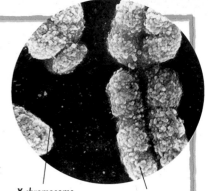

Y chromosome   X chromosome

The DNA in each cell is divided into short lengths called chromosomes. Humans have 23 pairs of chromosomes, 46 in total. Each chromosome is made up from a series of smaller units of DNA called genes. Some genes tell cells how to make the proteins that are essential for growth and the correct working of the cells, but scientists don't yet understand what most genes do. Not all the genes in every cell are switched on all the time. In muscle cells, for example, only the genes that make muscle proteins are switched on.

4.

3.

RNA

PAIRING UP

Chromosomes come in pairs because one copy is inherited from one parent and a second copy is inherited from the other parent. Each of a girl's chromosomes looks the same as its paired chromosome, but one pair of a boy's chromosomes look very different from each other. This pair contains one X chromosome and a smaller Y chromosome (*above*)

BOYS WILL BE BOYS?

All of a man's male characteristics are due to only one gene, called the SRY gene, found in the Y chromosome. This means a baby will be a girl unless the SRY gene changes it to a boy. It is as though the 'normal' state for human embryos is female unless the SRY gene is present. It works by switching on other genes that produce chemicals, including testosterone, that change the embryo's brain and body into a male.

SCIENCE EXPLAINED: CELL DIVISION

Cells don't last forever. Every hour, about 200 billion of your cells die and have to be replaced with new cells. Cells increase their numbers by duplicating and dividing. First, each chromosome in a cell duplicates to form two strands of DNA called chromatids. These chromatids are then pulled apart by fibres called spindles, and are shared between the two new nuclei before the two new cells are formed. The chromatids then become the new cell's chromosomes.

PASSING ON OUR GENES

ROYAL GENES

Genes can be traced through a royal family very easily because its family tree has been recorded accurately for hundreds of years. Queen Victoria (*above*) inherited a faulty gene that causes a condition called haemophilia. Someone who suffers from haemophilia cannot stop bleeding from a cut because the blood will not clot. Haemophilia affects only men, but the faulty gene is carried and passed on by women. Two of Victoria's daughters became carriers of the gene, and passed it on to the royal families of Russia and Spain.

We pass on our appearance and other characteristics to the next generation by giving them copies of some of our genes. Although some genes disappear through natural selection, many have been passed down from generation to generation for thousands of years. Scientists have worked out, for example, that the gene that produces ginger hair, fair skin and freckles in humans probably pre-dates modern man.

KEEP IT IN THE FAMILY

Inheritance allows genes to survive for generations. They are passed down through families, and the characteristics they produce can be traced through the branches of a family tree. New people join a family by marrying into it and introducing new genes that will be added to the existing mix of genetic material. Scientists use family trees to work out the likelihood of serious diseases like breast cancer being passed on to future generations.

Andrew Stevens (1897-1966)   Susan Kaplan (1905-1982)   Peter Smith (1860-1945)   Irene Roberts (1861-1953)

Lewis Stevens (1922-1980)   Robert Stevens (1923-1975)   Petra Stevens m. Maurice Smith (1928-) (1920-1984)   Rebecca Smith (1916-1990)

Darren Alexander m. Sarah Smith (1950-) (1954-)   Alex Smith (1960-)

Suzy Alexander (1979-)   Jack Alexander (1984-)

EXCEPTIONS TO THE RULE

The sex of most animals depends on the chromosomes they inherit from their parents. However, some reptiles such as crocodiles are different. The sex of a crocodile is not determined by its genetic code. Instead, a crocodile is born male or female according to the temperature of the egg it came from. In the case of the Australian saltwater crocodile, an egg temperature of 31–32° C produces males. If the eggs are a few degrees cooler or warmer then they hatch as females.

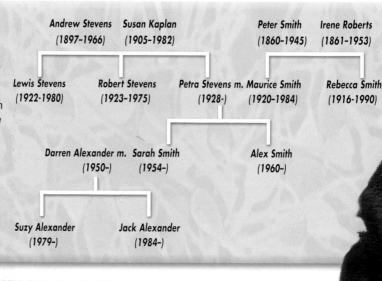

GINGER GENE

Some of our genes are very old. Scientists at Oxford University in England compared the gene for ginger hair in humans with a similar gene that appeared first in chimpanzees. They found 16 differences, or mutations, between the two. One of them produces ginger hair. They calculated that this mutated gene would have taken up to 100,000 years to pass through the generations and become as common as it is today. As modern man is only 40,000 years old, we probably inherited the gene from an older human species, possibly Neanderthal Man (left).

GREGOR MENDEL

The new science of genetics began with the work of a monk, Johann Gregor Mendel (1822–84). He grew thousands of pea plants, collecting the seeds and using them to grow new plants. He studied how often certain characteristics appeared in each generation. Even though he knew nothing about chromosomes, genes or DNA, Mendel worked out the basic laws of genetic inheritance – that offspring inherit their characteristics equally from both parents.

OUR GENETIC RELATIVES

Different species of plants and animals that are closely related to each other have a similar genetic code. We are closely related to chimpanzees and share up to 99 per cent of our DNA with them. Incredibly, it is only the last one per cent or so of our DNA that distinguishes us from chimpanzees.

SCIENCE EXPLAINED: DOMINANT & RECESSIVE GENES

We inherit two copies of each gene, one from each parent. They may be the same or different. When they are different, one gene may be dominant. A dominant gene hides the effect of a weaker recessive gene. In humans, for example, the gene for brown eyes is dominant, while the gene for blue eyes is recessive. A baby is born with blue eyes only if he or she inherits two copies of the blue eye gene, whereas only one brown eye gene is needed to give a baby brown eyes.

CASE STUDY: SOLVING MYSTERIES

CRIME SCENE

The scene of a crime today can sometimes look like a hospital operating theatre, with forensic scientists clothed in white from head to foot. The overalls ensure that the scientists do not drop fibres or biological material, such as skin flakes or hairs, onto the evidence. They must not allow evidence to be contaminated with their own DNA, because it could spoil the results of their tests. Outdoor crime scenes are sometimes covered by a tent to stop material being blown onto the area by the wind or washed away by rain. The image above is taken from the BBC drama *Silent Witness*, about the work of a forensic pathologist.

Today, many crimes and mysteries are solved using genetics. DNA left at the scene of a crime can be compared with DNA taken from suspects. If they are the same, this supports the case against a suspect — if they are different, the tests can prove that a suspect is innocent. DNA testing can also be used to clear up other controversies. One of the longest running mysteries surrounding a woman who claimed to be a member of the Russian Imperial family was finally solved by a simple genetic test.

BACK FROM THE DEAD?

During the Russian revolution of 1917, Tsar Nicholas II and his family were removed from power, and were reported to have been murdered by their guards. However, there were rumours that at least one of the children, possibly Anastasia, had survived. In the 1920s, a woman calling herself Anna Anderson came forward claiming to be Anastasia. Research into her past however suggested she was actually a Polish woman called Franziska Schanzkowska. After her death in 1984, genetic tests on a piece of her tissue and tissue from people related to the Russian Imperial family proved that Anna Anderson was not Anastasia. The 1956 movie *Anastasia* told the story of Anna Anderson, with the lead role played by Ingrid Bergman (*above*)

SCIENCE EXPLAINED: READING THE GENETIC CODE

Forensic scientists (scientists whose work is used in court as evidence) analyse biological samples to reveal their DNA. They use electrophoresis (see page 5) to analyse several samples of DNA at the same time. If the patterns of bands produced by the samples match exactly, then they probably came from the same person. If only some of the bands match, then the samples came from people who are probably related. If none of the bands match, then the samples are from people who are not related.

A GUILTY MAN?

Many samples of human tissue from crimes committed before genetic profiling was invented are still kept today. In crimes that were unsolved or where there is doubt about someone's guilt, these samples can be tested using new techniques. James Hanratty was hanged in 1962 for the murder of the scientist Michael Gregsten and the rape of Gregsten's mistress, Valerie Storie. Nearly 30 years later, after a long campaign from Hanratty's supporters to clear his name, DNA tests were ordered. These results were to prove a shock for Hanratty's family. They showed a match between DNA taken from Hanratty's body and traces on clothing found at the crime scene.

This is my son
JAMES HANRATTY
murdered by the state
for the A.6. murder

ELEVEN WITNESSES SWEAR HE
WAS IN RHYL 200 MILES AWAY
WHEN THE CRIME WAS COMMITTED

I DEMAND A
PUBLIC INQUIRY

AND JUSTICE
TO BE DONE

GENETIC PROFILING

In 1984, the British geneticist Alec Jeffreys (above) developed a new technique called DNA profiling, also known as genetic fingerprinting. He noticed that some sequences of DNA are repeated over and over again within a person's genetic code. He also found that, with the exception of identical twins, everyone has a unique pattern of these repeated sequences. He went on to develop a way of making them visible so that the patterns from several samples of DNA could be compared.

FISHING FOR GENES

The scientists above are looking at a magnified image of chromosomes that have been treated so that parts of them glow. The technique is called FISH, which stands for Fluorescence In Situ Hybridization. It is used to work out the base sequence of unknown DNA. Fish works by using short strands of DNA, called probes. When the probes are mixed with DNA, they plug into parts of the DNA that they fit. Scientists can see where they are because they have been treated with a chemical that makes them glow. And as the base sequence of the probes is known, that tells the scientists what the base sequence of the unknown DNA is. By using different probes that plug into different parts of the unknown DNA, the whole base sequence of the unknown DNA can be worked out.

SCIENCE EXPLAINED: AUTOMATIC SEQUENCING

Geneticists are developing new, faster ways of sequencing DNA. Electrophoresis (see page 5), the technique often used to analyse genetic material, involves passing an electric current through the DNA, which heats it up. But heat damages DNA, so early electrophoresis techniques had to work slowly to avoid a damaging build-up of heat. New techniques keep the DNA cool while using larger electric currents. The larger the current, the faster the process works.

CASE STUDY: THE HUMAN GENOME PROJECT

In 1988, scientists set out on the most important research project since the discovery of DNA. They started working out a map of the human genetic code, which is also called the human genome. The Human Genome Project involves mapping, or sequencing, human DNA provided by several anonymous donors. The resulting DNA map will be typical of all human DNA. Forming a map of human DNA is not the same as understanding what DNA does. It will take decades, maybe centuries, for scientists to work out what all of the many different genes do and how they do it.

LISTING BASES

The first stage in the Human Genome Project is to make a list of all the bases that make up the genes in human chromosomes (see *page 10-11*). This is done by superfast DNA sequencers (*above*), as there are about three billion bases in the human genetic code. Each short sequence of bases is worked out several times, so that any mistakes can be found and corrected. The next stage, which will be much more difficult, involves trying to understand exactly what the thousands upon thousands of human genes do.

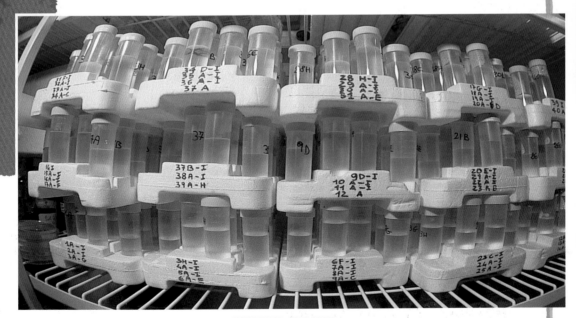

FINISHING THE MAP

The phials above contain every gene in the human body. Scientists drew up the first rough map of the human genome at the beginning of 2001. However, this first rough listing has lots of gaps in it that still have to be sequenced. The gaps are made from DNA which probably contains few genes. Scientists aim to fill in all of these gaps so that they have the complete DNA sequence of each chromosome from tip to tip by 2003.

SCIENCE EXPLAINED: MAKING A CLONE

One technique for making clones is called nuclear transfer. First, the nucleus is removed from an unfertilized egg. Next, a cell from an adult animal is fused with the egg by passing an electric current through the two. They become one cell which then behaves like a fertilized egg and begins to divide. Finally, the cell is implanted into a female where it develops normally into an embryo.

The adult sheep to be cloned.

Cells are removed from the adult sheep.

The new cell starts to divide like a normal cell.

One cell is fused with the egg.

The nucleus is removed from the egg.

An unfertilized egg.

The clone is born.

ORGANS TO ORDER

The first attempts to clone animals were made in the 1950s by inserting the nucleus of a frog cell into a frog egg, but these were not successful. Frogs were cloned successfully for the first time in 1970, but the tadpoles did not develop into adult frogs. To learn more about humans, scientists then turned their attention to cloning mammals. Five healthy piglets, Millie, Christa, Alexis, Carrel and Dotcom were cloned from adult cells and born on March 5, 2000. The hope is that one day such pigs with modified cells can be cloned to provide organs for human transplantation.

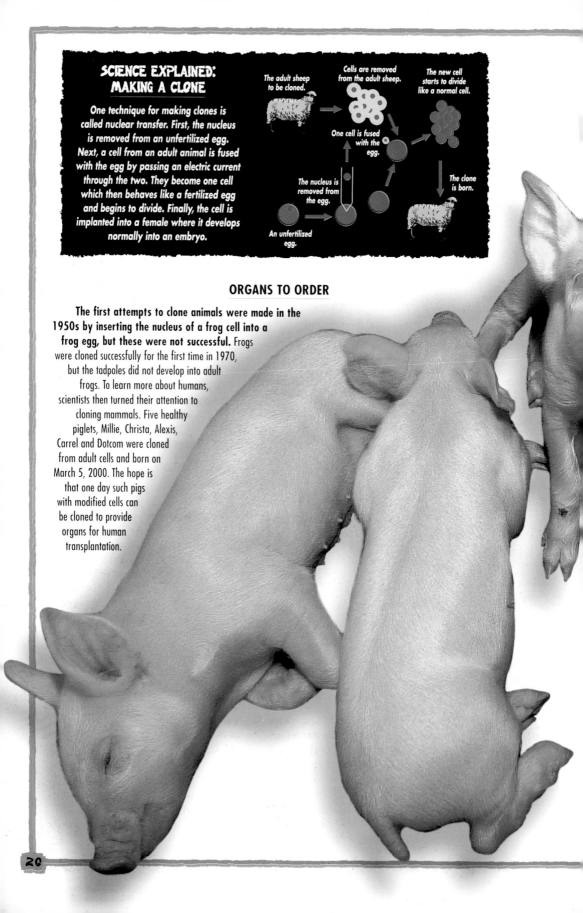

CASE STUDY: CLONING

Cells and whole plants and animals that are genetically identical to each other are called clones. Clones can occur naturally (identical twins, triplets and so on are natural human clones), but they can also be made in the laboratory. Making clones is risky, because many do not develop at all and some seem to have genes that don't work properly. Clones are very useful in scientific research. They enable scientists to test the effects of different chemicals, drugs and medicines on identical cells and organisms. It is possible to make human clones, but most people think that scientists should not make people by cloning them in laboratories.

DOLLY THE SHEEP

Dolly the sheep made headlines worldwide in 1996, because she was the first mammal to be cloned from an adult cell. Animals had already been cloned using embryo cells, as they can develop into all the different types of cells a new creature needs. However, adult cells become set in their ways. Dolly's creators found a way of making the adult cell behave like an embryo cell to produce all the different cells that Dolly needed.

IDENTICAL TWINS

Making clones in a laboratory is one way of producing genetically identical animals, but they also occur naturally, in about one in every 80 human births. When one fertilized egg splits in two, identical twins are born. Roughly one birth in every 6,400 results in triplets, three identical babies developing from the same fertilized egg. Four identical babies occur only once in every half a million or so births; the chances of having five identical babies is just one in 41 million.

ANIMAL BREEDING

People have been changing animal genes for thousands of years by selective breeding. Choosing which animals to breed from reinforces some genes and eliminates others. Farmers use selective breeding to improve certain qualities in their animals. Today, people like less fatty meat, so cattle and pigs have been bred to produce leaner cuts. Geneticists are also playing the role of conservationists, helping to ensure many endangered species continue to thrive.

A MEATY ISSUE

Up to about 300 years ago, cattle breeds developed naturally in different places. But from the 18th century onwards, selective breeding produced a variety of new breeds. The first of these was probably produced by crossing white-faced Dutch cattle with small black Celtic cattle to create the meaty Hereford breed.
At first, the cattle were bred naturally by putting selected bulls and cows together. Then, in the early 1900s in Russia, a technique called artificial insemination was introduced. Today, it enables one bull to produce up to 10,000 calves.

BREEDING WINNERS

The breeding of animals used for racing, such as dogs and horses, is strictly controlled and recorded, generation after generation. Champions are in great demand for breeding, because their offspring have the best chance of being champions themselves. Thoroughbred horses often have a DNA profile made and stored in an official register. Another DNA profile can be made at any time and compared to the original profile to prove a horse's identity or to prove which horses it was bred from.

LIFE SAVER

Endangered animals can be saved from extinction by breeding projects aimed at increasing their numbers. A good example of this is the giant panda. There may be only 1,000 of them living wild and perhaps another 120 in zoos. The first giant panda bred using artificial insemination was born in 1963 at the Chengdu Giant Panda Breeding and Research Centre in China. The centre has now produced 34 giant pandas by the same method.

DOGS & CATS

All the breeds of dogs that we have today were produced artificially by selective breeding. The first dogs evolved from prehistoric wolves and jackals. Up to 6,500 years ago, there were still probably only five types of dog – mastiffs, wolf-type dogs, sight hounds (such as greyhounds), pointing dogs and herding dogs. Selective breeding between these types has produced all the modern breeds.

DNA DETECTIVE

It is illegal in some countries to take hawks or their eggs from the wild. If the authorities are suspicious about a hawk owner's claim that a bird was bred in captivity, DNA profiling (*see page 17*) can prove or disprove his claim. If the bird was bred in captivity, its DNA will closely match the DNA of the birds from which it was bred, but if it was taken from the wild, their DNA will not match. In 1993, a hawk breeder in Scotland was convicted of nest-theft, proven by DNA profiling.

SUPERPLANTS

Genetic engineering can produce new varieties of food plants such as tomatoes that stay ripe for longer without rotting. It can also produce new varieties of crops that are more resistant to disease. This would prevent disasters such as the Irish potato famine, which occurred in the 1840s, during which more than a million people starved and even more left the country for a new life in North America. The famine had been caused by the failure of the entire potato crop because of a fungus disease called potato blight.

REMARKABLE RICE

About four billion people depend on rice, so it is important that rice crops do not fail. Rice was the first major food crop to have its genome mapped. Scientists are now modifying its genome to develop new varieties of 'super-rice'. Low levels of vitamin A and iron in many people's diets in the Third World lead to blindness and anaemia, so scientists have used genetic engineering to produce a variety of rice that has more vitamin A and iron.

SCIENCE EXPLAINED: GETTING NEW GENES INTO PLANTS

There are several ways of getting new genes into plants. Bacteria known to infect the plants can be used to carry the gene DNA. Alternatively, the genes can be stuck to metal particles that are then fired into the plant cells like bullets. Other methods include using electric shocks to break down plant cell walls to let in new genes, or injecting new genes directly into the plant cells through a needle.

USING BACTERIA

DNA to be inserted.

USING GENE BULLETS

Bacteria carries the DNA in a ring called a plasmid.

Metal particles are coated with DNA.

Plant cells

The bacteria attacks the plant cells, 'infecting' it with the DNA.

The particles are fired into the plant cells.

In soil, a plantlet grows into a genetically adapted plant.

The altered plant cell begins to divide.

Cells are regenerated into plantlets.

CHANGING PLANTS

The crops we eat today were developed from smaller wild plants by selective breeding. The biggest and best plants were used to pollinate each other, generation after generation, to strengthen the features the growers wanted, such as the size of the fruit. Geneticists are now speeding up the process by changing plant DNA in the laboratory. This is called genetic engineering. Scientists use genetic engineering to change plants such as barley and maize so that they can better resist diseases, pests and drought conditions and grow in a wider range of soil types. They are also producing fruit that lasts longer without rotting. Crops that have been altered genetically are known as genetically modified or GM crops.

WHEAT TO EAT

Grains like wheat are important food crops. A large part of the world's agriculture relies on only five types of grain. It is therefore important to map and understand their genomes so that they can be modified to resist being wiped out by pests, disease or a changing climate. Mapping grain genomes will also enable scientists to develop more resilient strains of wheat in the future.

THE GREAT GENE HUNT

Scientists are continually checking plants from all over the world for genes that might be used to modify another plant. Any plant might have a gene that could improve the properties of another one. It is particularly important to save rare plants from extinction, otherwise any useful genes they contain could be lost forever.

GENETIC ENGINEERING

GERMLINE THERAPY

Most gene therapy affects only the patient who has been treated. The treatment does not affect the sex cells so it is not passed on to any children the patient may have. In 1999, a team of Canadian scientists succeeded in giving a mouse an extra chromosome that was passed on to the mouse's offspring. If the same technique were to be used in humans, any genetic change would be passed on to future generations. This technique is called germline therapy.

Scientists can now not only read the genetic code inside human cells, they can actually change it by genetic engineering. About 3,000 diseases and damaging medical conditions in humans are caused by inheriting faulty genes. Treating them by using genetic engineering is called gene therapy. The 19th century was dominated by engineers who worked with iron and steel. The 20th century was dominated by electronic engineers who developed the computers and miniature electronic equipment that we use now. The 21st century may be dominated by the work of genetic engineers and gene therapists.

SCIENCE EXPLAINED: ALTERING VIRUSES

A virus that is to be used to carry a gene into a cell is first made safe by removing some of its genetic material. A human gene is then introduced into the virus, where it becomes part of the virus's own genetic code. Trillions of modified viruses are injected into the patient. Each virus attaches itself to a cell and sends its genetic material into the cell where it becomes part of the cell's DNA.

The virus with its genetic material removed.

The virus attaches to the patient's cell and empties genetic material into it.

THE ENGINEERED GENE

The virus takes up the gene in its own genetic code.

THE PATIENT'S CELL

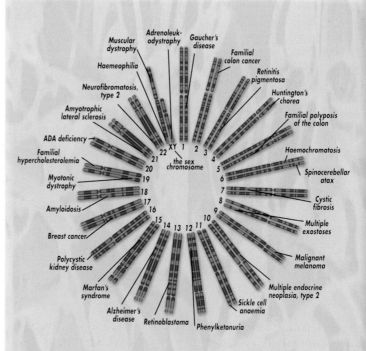

GENETIC DISEASES

Geneticists have found the locations of genes that can cause many of the inherited conditions that people suffer from. Above is a complete set of the 23 pairs of human chromosomes, showing in which genes these diseases occur. Geneticists located them by studying families where several relatives suffered from the same condition. Their chromosomes were compared to those taken from healthy families. The differences showed where the faulty genes were located.

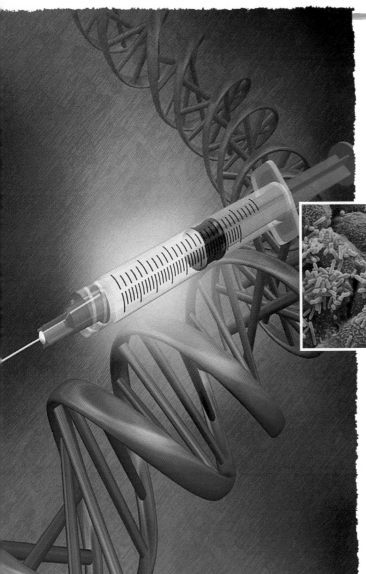

GETTING IN

New DNA can be injected directly into a cell using a thin needle, but there are other better methods. The cells can be treated with chemicals that make their cell membranes 'leaky', so that the new genes can pass through into the nucleus. However, the most common methods of getting DNA into cells involve using viruses and bacteria.

BACTERIA BOON

Bacteria are used as microscopic factories to produce all sorts of chemicals to correct faults in the body. Using genetic engineering, a gene that produces the chemical is added to the bacteria's own genes. Chemicals such as hormones, insulin (for treating diabetes), and a variety of proteins are all produced in this way. Without genetically engineered bacteria, these substances would have to be extracted from other people, which involves a risk of infectious illnesses spreading from the donor to the patient.

CYSTIC FIBROSIS

ystic fibrosis is the most common inherited illness. It affects mainly the lungs, making it difficult to breathe.
s caused by a single faulty gene on chromosome 7. The gene is recessive, so two faulty copies must be inherited –
ne from each parent – to cause the illness. It is most common in Europeans, where it affects one person in 2,000.
cause cystic fibrosis involves only one gene, it is potentially one of the simplest diseases to treat genetically. Once
a system has been developed to deliver a healthy gene, the disease should be conquered. Scientists are
experimenting with treatments that involve using viruses to carry new, healthy genes directly into the lungs.

VIRUS CARRIERS

Viruses like the adenovirus (*left*) are very good at getting inside cells and combining their genetic material with material that already exists. They are ideal 'carriers' to take engineered genes into cells. The adenovirus, which normally causes a cold-like infection, is often used for this purpose.

DESIGNER BABIES

Parents want the best for their children. As new genetic techniques are developed, it will probably be possible to select certain characteristics for children. People who can pay will ensure that their children have the best genetic make-up to give them the best chance in life, while others won't be able to afford this luxury. This could result in a genetic underclass of people with more faulty genes than richer people.

genetiX snowball

0161 834 0295

SCIENCE EXPLAINED: A MIXED BLESSING

Blood cells are normally disc shaped, but people who suffer from an inherited condition called sickle cell anaemia have red blood cells shaped like a sickle or a crescent. Two copies of the faulty gene must be inherited to cause the illness. Gene therapy could wipe out the faulty gene. That seems to be a good thing, but scientists believe that inheriting one copy of the faulty gene provides protection against malaria. Wiping out the sickle cell gene could mean that otherwise healthy people would die from malaria.

ETHICS OF GENETIC ENGINEERING

Genetic engineering can improve crops, create more productive farm animals and cure diseases, but many people are very worried about its use. They are concerned that altered genes may behave in unpredictable ways once they are released from laboratories and spread into the environment. Every new genetic discovery and technique raises more questions about the ethics of genetic engineering. Some people worry that because genetic engineering is such a new technology, we do not know its long-term implications, and the risk of a terrible disaster is too high.

GENETIC TESTS?

There are now genetic tests that can detect a variety of inherited illnesses including muscular dystrophy, cystic fibrosis and haemophilia. If your family is affected by an inherited illness and there is a test that can show if you have inherited the faulty gene, should you have the test? Would you really want to know that your child may develop an illness for which there may be no treatment? Many people think genetic tests pose serious ethical and religious problems.

GM CROPS: GOOD OR BAD?

Activists often destroy GM crops, not agreeing with scientists who feel they are necessary to help feed people in the Third World. Some say that existing Third World crops could be made more productive by improving the soil, dealing with insect pests, and persuading Third World governments to spend less on arms and more on agriculture.

GENETIC DISASTERS

Scientists were reminded that gene therapy carries risks like any other treatment when a patient they were treating died in 1999. Jesse Gelsinger, an 18-year-old man from Arizona, suffered from a rare inherited liver disorder called OTC deficiency. He was being treated with adenoviruses altered to carry healthy OTC genes into his liver. When trillions of the viruses were injected into his body, he suffered a violent reaction to them and died. It is worth remembering that, to date, no-one has been cured of any disease using gene therapy.

GENETICS TODAY & TOMORROW

We are at the very beginning of genetic research and engineering. No-one knows how it will affect our species in the next hundred years, but our new knowledge of how our bodies work seems certain to completely change our understanding of diseases and how to treat them. Genetic engineering and gene therapy hold the promise of cures for many diseases including cancer, an end to the famines that kill millions of people, and possibly an end to the random lottery of natural evolution. It could be a scientific and medical revolution on a truly amazing scale.

GROWING CORNEAS

The cornea is the transparent 'window' at the front of the eye. It is actually a complicated structure with three layers of cells. In 2000, scientists succeeded in growing an artificial cornea that worked like a real one. The correct genes were switched on in the correct layers of cells. The artificial cornea will be used in testing the safety of medicines, which will reduce the number of tests done on animals.

THE BODY SHOP

Stem cells are found in bone marrow and in embryos. They are the 'general-purpose' cells that multiply easily and develop into specialized cells such as skin, blood, muscle or liver. Scientists are hoping to eventually be able to use them to produce human organs virtually on demand. Modified stem cells pass on their new genes to every new cell, but scientists don't yet know how to make them change into all the different types of cell they need.

WALKING AGAIN?

Our bodies are very good at repairing damage. Cut skin and broken bones heal themselves quickly. However, a damaged spinal cord will not heal itself. Someone whose spinal cord is severed will be paralysed. In future, scientists may be able to switch on the gene for spinal cord cell growth and enable paralysed people to walk again.

AN OIL FIELD?

The fuel we burn in our cars is made from oil. The oil is pumped out of underground oil fields, but one day these oil fields will run dry. In future, we may be driving cars powered by plant fuel. Plants such as sunflowers and rape may be genetically modified to produce an oil that is as effective as the fuel we use today.

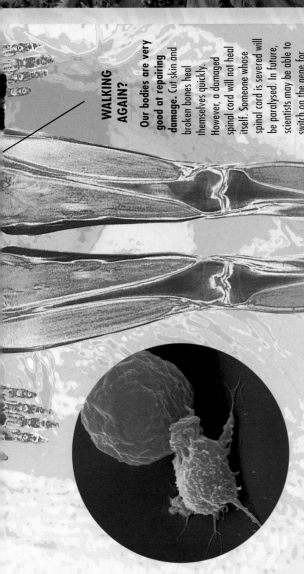

KILLING THE KILLER

Cancer is caused by cells multiplying out of control. They form a growth called a tumour (*see above left*). Tumour cells that are carried around the body by the blood system can start new tumours growing in other places. Because the cells belong to the body and are not alien invaders, the body's immune system does not attack them. Some cancers are caused by a single faulty gene, so scientists think they can be dealt with by using gene therapy to correct the fault.

GLOSSARY

Amino acids – Simple chemical compounds that join together inside cells to make proteins

Cell – The smallest unit of life that can function on its own

Chromatid – A thread of DNA formed during cell division when a chromosome splits along its length

Chromosome – A short length of DNA inside a cell's nucleus

Clones – Genetically identical organisms

Cytoplasm – The jelly-like material surrounding the nucleus inside a cell

DNA – Deoxyribonucleic acid. The genetic material inside cells

Dominant gene – A gene that produces a characteristic in an organism when only one copy of the gene is present

Double helix – The shape of the DNA molecule

Endoplasmic reticulum – A network of folded membranes in a cell where proteins are made

Gene – The smallest part of a chromosome that determines a characteristic of an organism

Genome – All the genes in a set of chromosomes

Golgi complex – An object inside an animal cell that stores and distributes chemicals

Inheritance – Characteristics passed on from the previous generation

Mitochondria – Objects inside cells that release energy to power chemical reactions

Mutation – A change in a cell's DNA

Nucleus – The control centre of a living cell, containing most of its DNA

Protein – A molecule in the shape of a long chain of chemical compounds called amino acids found in living cells

Protoplasm – The material that fills a cell

Recessive gene – A gene that produces a characteristic in a plant or animal only when two copies of it are present

Ribosome – A ball of RNA inside a cell that makes proteins according to instructions from the DNA in the nucleus

SRY gene – The gene in a human Y chromosome that makes a baby male

Virus – A particle of nucleic acid (a part of DNA) that can multiply only when it is inside a living cell

ACKNOWLEDGEMENTS

We would like to thank Advocate, Oliver Zaccheo from the Department of Genetics at the University of Leicester and Elizabeth Wiggans for their assistance.
Illustrations by John Alston and Simon Mendez.
Copyright © 2001 ticktock Publishing Ltd.
First published in Great Britain by ticktock Publishing Ltd., Century Place, Lamberts Road, Tunbridge Wells, Kent TN2 3EH, U.K. All rights reserved.
No part of this publication may be reproduced, stored in a retrieval system, or transmitted in any form or by any means electronic, mechanical,
photocopying, recording or otherwise, without prior written permission of the copyright owner.
A CIP catalogue record for this book is available from the British Library. ISBN 1 86007 259 3 (paperback). ISBN 1 86007 263 1 (hardback). 1 2 3 4 5 6 7 8 9 10
Printed in Malta

t=top, b=bottom, c=centre, l=left, r=right, OFC=outside front cover, IFC=inside front cover, IBC=inside back cover, OBC=outside back cover

BBC Photo Library: 16tl. Corbis Images: IFC, 2c, 3b, 8tl, 17r, 21br, 29tr. Corbis Stockmarket: 28t. Environmental Images: 31t. Kobal Collection: 16-17c. Mary Evans Picture Library: 14tl.
Pictor: 10-11t. Popperfoto: 20-21c. Science Photo Library: OFC (main), 2tl, 3t, 4d, 4b, 4-5c, 5t, 6b, 7t, 7b, 9t, 9r, 10b, 11t, 11c, 12b, 13t, 14t, 14b, 15b, 16b, 18t, 19t, 19b, 21cr, 22t,
22c, 22br, 23cl, 23r, 24l, 24c, 25r, 25b, 26t, 27t, 27r, 30-31c, 31b. Tony Stone Images: OFC, OFC (inset), 6l, 6-7c & OBC, 14–15b & OBC. Still Pictures: 28-29.

Every effort has been made to trace the copyright holders and we apologize in advance for any unintentional omissions.
We would be pleased to insert the appropriate acknowledgement in any subsequent edition of this publication.

snapping-turtle
guide